Center for Hydrogen Safety Conference Europe 2023

AA008756

Rotterdam Ahoy, Netherlands
9-11 May 2023

ISBN: 978-1-7138-8676-1

Printed from e-media with permission by:

Curran Associates, Inc.
57 Morehouse Lane
Red Hook, NY 12571

Some format issues inherent in the e-media version may also appear in this print version.

Copyright© (2023) by AIChE
All rights reserved.

Printed with permission by Curran Associates, Inc. (2024)

For permission requests, please contact AIChE
at the address below.

AIChE
120 Wall Street, FL 23
New York, NY 10005-4020

Phone: (800) 242-4363
Fax: (203) 775-5177

www.aiche.org

Additional copies of this publication are available from:

Curran Associates, Inc.
57 Morehouse Lane
Red Hook, NY 12571 USA
Phone: 845-758-0400
Fax: 845-758-2633
Email: curran@proceedings.com
Web: www.proceedings.com

TABLE OF CONTENTS

System Integration North Sea Introduction of Hydrogen as Energy Carrier in Offshore Energy Production and Exploration 1
Pim Reuderink

Changing Landscape / Evolution of the Hydrogen Ecosystem 2
Andrei V Tchouvelev

Inherently Safe Designs for Hydrogen Use 3
Josué Quilliou, Maxime Duchateau, Sara Jeghloul, Yanis Bouchemoua

Negligible Extent Classification and Specific Detailed Risk Assessment in a Hydrogen RefuellingStation (HRS) Example 4
David Torrado

Navigating Unfamiliar Territory—US Hydrogen Safety Regulations and European Standards 5
Achim Wechsung, Alfonso Ibarreta, Timothy Myers

Pinpointing Hydrogen Leaks from a Safe Distance with Ultrasound Cameras 6
Harmina Vejayan, Florian Perrodin

Hydrogen Detection – Sensor Technology to Enable Improved Safety and More Reliable Hydrogen Applications........ 7
Martina Wennesjö

Assessing Differences of Ignited H2 and Hydrocarbon Vent Releases: Design Considerations and Safety Risks........ 8
João Vitor Assunção, Vinicius Esteves, Bruno Wehmuth, Rafael Pereira

Process Safety Leadership in the Emerging Hydrogen Sector 9
Trish Kerin

Lessons Learned from a Fire at a Hydrogen 200 Bar Filling Station 10
Sidonie Ruban

Hydrogen Safety in Rail Applications 11
Enrico Morelli

IEA TCP Task 43- Subtask Safety Distances: State of the Art 12
Elena Vyazmina, Lee Phillips, Richard Chang

The Hygrid Project - A Hydrogen Blending Demonstration in a Gas Distribution Network - Design Considerations for Safe Operation 13
Pradheep Kileti, Christopher Cavanagh, Tara Schneider-Moran, Ron Bevacqua, John Wolf

Comparison of a Prescriptive and a Performance-Based Approach for Fire and Explosion Risk Assessment of Hydrogen Plant........ 14
Xavier Deckers, Khodadad Kamaliseresht

Engineering Assessment Process for Hydrogen Blending in Natural Gas Transmission 15
Bradley Vonau

Liquid Hydrogen Maritime Transportation 16
Julien Boulland

Refining Hydrogen VCE Modeling Approaches 17
Peter Diakow, Darren Malik

Hydrogen Safety Management – Essential Practices for Risk Management in Operational Facilities 18
Venkata Sainath Garimella, Raman Swaminathan, Senthilraja Mani, Nagarajan Ramachandran, Raja Dandamudi

Hydrogen Use for Decarbonized Aircrafts and Associated Safety Challenges 19
Pierre Wolf, Beatrice Toussaint, Alban Mahon, Anaïs Marie Gay

Process Safety Management for the Hydrogen Industry 20
David Moore

Leak Detection for Hydrogen Gas: Atmospheric Dispersion and Optimal Sensor Placement 22
Rakesh Paleja

Evaluation of Safety Distances for Hydrogen Refueling Stations 23
Benjamin Truchot, J. Daubech

Multhyfuel Project for a Safe Implementation of Hydrogen Refuelling Station in a Multifuel Context : Midterm Achievements 24
Quesnel Sébastien

Safety Considerations in Cryogenic Liquid Hydrogen Bunkering 25
Jennifer Wen

Safety-Related Activities in the Clean Hydrogen Partnership 26
Alberto Garcia Hombrados

Mitigation of Electrostatic Ignition Hazards During the Bulk Loading and Fueling of Road Tankers, Railcars and Public Transportation Vehicles. 27
Mike O'Brien, Jonathan Rowe

Eliminating Hydrogen Fire in CCR Units 28
Manea Alharkan, Majdi Almutairi

Safe Measures to Mitigate the Effects of a Surrounding Fire on Gaseous Hydrogen Composite Tube Trailers 29
Christophe Guerin

About Hydrogen Leak Detection. Overview of Existing Devices and Investigations on Innovative Technologies for Safety Considerations. 30
Deborah Houssin

Author Index

System Integration North Sea introduction of Hydrogen As Energy Carrier in Offshore Energy Production and Exploration

Tuesday, May 9, 2023 - 3:55pm to 4:10pm

Authors

Reuderink, P. - Presenter

Introduction

At the moment, in the North Sea, new tender rules are implemented. The focus is instead of building a wind farm with one energy carrier, to have at least two energy carriersâ€™ electrons and molecules.

What is considered, is the main high voltage grid in the Netherlands is congested. The landing to shore of the wind power is under development, to speed up the energy transition, molecules could be an intermediate solution.

Integrated North Sea energy systems

To have an integrated system on the North Sea, export by High Voltage cables and at least one other energy carrier is needed. There can be central production of hydrogen by an offshore structure with an electrolyser including rectifiers and water treatment systems near an offshore substation or by decentral hydrogen production on the base of the wind turbine with interfiled pipeline connection to an export pipeline that centralise landing to shore.

Another added value is that (seasonal) storage can be done offshore in depleted gas-fields or new built salt cavern storage subsea.

The RVO (Netherlands Enterprise Agency) has pointed out based on decision made in 2022 of the policy makers of the Dutch ministry Climate and Energy transmission, storage and distribution of various energy carriers, such as electricity and gas (including hydrogen gas) need to be adopted in any future offshore energy projects.

System approach

The following key aspects of an integrated system approach are high added value:

- Balance of grid â€“ efficient export of energy to shore
- Offshore storage solutions
- Not only limited to offshore wind, also solar, tidal, waves and so on can be considered a in new projects.
- Optimised sea and seabed use for the North Sea.
- Solution to speed up transmission of energy to shore and interconnectors to other countries.
- Lower environmental impact for offshore energy projects.

Changing Landscape / Evolution of the Hydrogen Ecosystem

Wednesday, May 10, 2023 - 9:20am to 9:50am

Authors

Tchouvelev, A. V. - Presenter

Hydrogen ecosystem is evolving due to rapid and significant changes that hydrogen energy applications bring to a traditional industrial hydrogen landscape. After a quick overview of the Hydrogen Council, the presentation addresses key evolution features as well as the uniqueness of hydrogen ecosystem vs natural gas. Examples also include road mobility and sustainability attributes.

Inherently Safe Designs for Hydrogen Use

Thursday, May 11, 2023 - 10:20am to 10:35am

Authors

Quilliou, J. - Presenter

Duchateau, M.

Jeghloul, S.

Bouchemoua, Y.

The hydrogen industry is developing fast and safety experts have to keep up the pace integrating hydrogen specificities. A sensible approach is to adapt practices that has already been developed for other industries. However, time is of the essence when dealing with hydrogen safety cases, therefore incorporating effective mitigation strategies and inherent safety concepts at the start of the project can avoid surprises and costly re-designs later on. This paper will present a number of strategies based on Gexconâ€™s experience within the hydrogen industry, which have been quantified and illustrated thanks to the use of CFD modelling.

Early isolation is key to limit the released quantity of hydrogen which could potentially lead to a vapour cloud explosion in case of delayed ignition. Relying on a detection system is thus a requisite but is sometimes not sufficient to prevent hazardous conditions. It is thus necessary to implement safety measures that follow simple concepts making the equipment inherently safer.

The first concept deals with the largest inventory, namely the hydrogen storage. The use of automatic shut-off valves right at the bottles would instantly stop the hydrogen feeding to the system.

The second notion relates to the size of the available free volume. The larger it is, the larger the hydrogen cloud might become. Partitioning to limit the free volume before releasing to a vent mast is proved valuable and effective.

The third proposition relies on hydrogen natural high buoyancy. Considering sloping ceiling would allow rapid evacuation of any release and avoid any accumulation. Different designs present different features which could be relevant depending on the use and restrictions (space etc).

When considered early on, inherently safe design would benefit the whole project without the need to investigate and invest in sophisticated equipment or procedure.

Negligible Extent classification and specific detailed risk assessment in a Hydrogen RefuellingStation (HRS) example

Tuesday, May 9, 2023 - 11:50am to 12:05pm

Authors

Torrado, D. - Presenter

The MultHyFuel project aims to develop evidence-based and harmonised guidance for the safe deployment of Hydrogen Refuelling stations (HRS) in a multi-fuel context. In order to generate good practice guidelines, the project has developed risk assessments using three HRS configurations for the study and analysed the Hazardous Area Classification (HAC) for those examples. Firstly, the hazardous area classification has been analysed for a representative dispenser with operating pressures between

350 and 875 barg. The background concentration and release rate has been studied at different pressures and natural ventilation rates. The results suggest that Negligible Extent (NE) could not be applied for the dispenser example due to the operating pressures of that stage of the system. Although this classification would not apply for the dispenser, it has been anticipated that that NE classifications may be applied for other sections of the HRS, for instance, a hydrogen generator. Therefore, an example of the NE classification and specific detailed risk assessment requirement as per IEC 60079-10-1:2020 for an enclosure within a hydrogen generator is presented. Recommendations for the specific detailed risk assessment are proposed for this scenario with the intention to support improved definition of the requirement in IEC 60079-10-1:2020 for future revisions.

Navigating Unfamiliar Territory—US Hydrogen Safety Regulations and European Standards

Tuesday, May 9, 2023 - 10:30am to 10:45am

Authors

Wechsung, A. - Presenter, Exponent

Ibarreta, A., Exponent Inc

Myers, T., Exponent, Inc.

Developers of hydrogen technology projects are interested in reaching as many customers as possible. The EUâ€™s hydrogen strategy and REPowerEU projects, as well as the funding available through the US governmentâ€™s Inflation Reduction Act are expected to boost hydrogen-related projects on both sides of the Atlantic. However, manufacturers, project developers and engineers need to be aware of local safety standards and regulation as they bring their products to the market.

European manufactures, developers and engineers looking to reach US customers may face an unfamiliar regulatory and standards regime. In our presentation, we will provide a brief overview of key US government regulators for safe handling of hydrogen. We will also review relevant US standard organizations with which Europeans designers may not be familiar, including ASME, NFPA, and CGA.

A more in-depth case study compares the European and US standards for the design of deflagration vents, EN 14994 and NFPA 68. Deflagration vents are an established technique for preventing catastrophic explosions and enclosure failures after a flammable fuel-air mixture is ignited. Sizing these devices for hydrogen presents a unique challenge due to hydrogenâ€™s combustion properties including a fundamental burning velocity that is almost 10 times higher than more conventional hydrocarbon fuels. Burning velocity and K_G (deflagration index for gases) are important parameters when sizing deflagration vents. The high burning velocity, and corresponding high K_G, of hydrogen can limit the effectiveness of explosion vents to mitigate hydrogen explosions. In this paper, we will compare these different approaches for sizing such devices, namely NFPA 68 and EN 14994 for several case studies.

Pinpointing hydrogen leaks from a safe distance with ultrasound cameras

Tuesday, May 9, 2023 - 2:00pm to 2:20pm

Authors

Perrodin, F. - Presenter

Vejayan, H.

While hydrogen is enjoying an unprecedented momentum, its production, storage and transportation pose technical challenges, as hydrogen is highly flammable. In particular, leak detection method is an essential step to achieve hydrogen safe use. However, traditional hydrogen leak detection methods pose some limitations. On the one hand, fixed hydrogen concentration detectors can only detect if a leak is present, without revealing the location of the leak. Moreover, these detectors have high chances of missing leaks when outdoors. On the other hand, portable sniffers are tedious to use, and can put operators at risk by being within potentially explosive atmospheres. In addition, leak findings strongly depend on the operator and how thorough inspections are made.

Here we propose ultrasound cameras for hydrogen leak detection, the unique method to pinpoint leak locations from a distance, without contact. The Distran ultrasound camera detects leaks by detecting the ultrasounds they emit. When the ultrasound waves emitted by a leak reach the front panel of the camera, the sensor is able to triangulate the origin of the ultrasound emissions, being the leak point. In real-time, the ultrasound camera provides a clear pinpointing of the leak on an wide angle optical image of the surroundings, enabling operators to precisely know the source of the problem and therefore act quickly. Besides leak detection, the Distran device also provides an option to quantify leak rates (in L/h), by using a specific quantification model that converts the sound pressure level emitted by hydrogen leaks into volumetric leak rates, enabling the classification of leaks to stablish maintenance priorities. The technology is also efficient outdoors, as well as within noisy industrial environments, as the sensor filters out noises outside of the ultrasound detection bandwidth. In addition, Distran offers an intrinsically safe ultrasound camera, Ultra Pro X, which can therefore enter hazardous areas.

Through real life examples, including inspection pictures and videos, Distran will demonstrate the capacity of ultrasounds cameras to pinpoint hydrogen leaks from a distance, and quantify their leak rates to coordinate relevant maintenance operations.

Hydrogen Detection – Sensor Technology to Enable Improved Safety and More Reliable Hydrogen Applications.

Thursday, May 11, 2023 - 10:05am to 10:20am

Authors

Wennesjö, M. - Presenter

The Insplorion sensor platform for hydrogen concentration measurements enables easy implementation and high reliability. The principle of operation is NanoPlasmonic Sensing (NPS), an optical, real-time, label free, surface sensitive sensor technology. The sensor chips consist of a Pd-based sensing element with optical read-out, which has been demonstrated to provide fast response with high selectivity for hydrogen.

Nano-sensing offers the capability to fine-tune the dynamics of gas interaction, which has demonstrated the possibility to sustain a response time less than a second. As time is of essence when it comes to hydrogen safety, seconds could prevent severe and costly accidents.

The sensing element does not require any particular background for operation, which means it can be used in air, inert atmosphere and even vacuum. Further, the pd-based sensor chip is inherently selective to hydrogen which means it is able to be used in environments involving other interfering gases. This enables robust and accurate monitoring for several new hydrogen application areas where current sensors on the market cannot function.

Currently, the sensor is in commercial prototype phase and Insplorion is proving technical features together with customers in pilot projects.

Assessing Differences of Ignited H2 and Hydrocarbon Vent Releases: Design Considerations and Safety Risks

Tuesday, May 9, 2023 - 4:10pm to 4:25pm

Authors

Esteves, V. - Presenter, DNV

Assunção, J. V.

Wehmuth, B.

Pereira, R., DNV

Hydrogen has been increasingly studied and applied in the energy and maritime industries, whether for heating, electricity, or as a fuel itself. However, in the absence of specific requirements for the design of ships using hydrogen as a fuel, the alternative for these projects is to demonstrate the safety level equivalence with ships using conventional fuels, the so-called â€œAlternative Designâ€� approach. With the expanded applications and the already well consolidated knowledge from the oil and gas industry, some hazards might be underpredicted. The current paper studies the different vent requirements for LNG and hydrogen gases and the potential risks associated in case of an accidental ignition leading to a jet fire. The radiation modeling is assessed using DNV PHAST software, utilizing the recently implemented Miller formulation. The consequence modeling performed of the radiated heat indicates that the minimum vent height requirement is deemed one of the most critical parameters one should take into consideration for the hydrogen vents design. The radiation levels can pose severe hazards to personnel close to an ignited hydrogen vent designed with the inappropriate height. The findings also indicate that there is a significant difference in the heat flux to which one could be exposed for these two distinct gases; for LNG, the lower heat of combustion plays an important role to justify the less intense radiation magnitude. In conclusion, although experiences from the traditional hydrocarbon industry can be greatly useful for hydrogen projects, it is critical to keep in mind the differences between these chemical compounds and properly assess the risks, in terms of frequency and consequences, of the application of hydrogen, including the accidental scenario of an ignited vent leading to a jet fire.

Process Safety Leadership in the Emerging Hydrogen Sector

Thursday, May 11, 2023 - 9:05am to 9:35am

Authors

Kerin, T. - Presenter, IChemE

As the world continues the transition to net zero, a number of new, and not so new technologies are being implemented at scale. The application of hydrogen is one such technology. From the traditional chemical and oil and gas sectors, both hydrogen is a well known and used substance. There is a clear understanding of how to design, deploy and maintain the technology. Many companies in these traditional sectors are expanding their application and use of hydrogen, but there are a number of new entrants into this sector. There are substantial hazards in dealing with hydrogen, and this needs to be managed to a tolerable risk level, to ensure that this fuel source remain a viable option in the energy transition and beyond. The traditional sectors have long established process safety systems and leadership activities, but the newer entrants may not have this same level of experience. The endite life cycle of hydrogen must be included in the safety management and use, from hazards associated with manufacturing, through the supply chain and end use applications. Regardless of the life cycle stage an organisation may be working in, leadership is an important element in delivering the necessary culture for safe outcomes. This presentation will look at what is needed from a process safety leadership and culture perspective for the safe manufacture, transport, handling and use of hydrogen for those new to and existing in ,the sector, as well as examine process safety implications of new technologies in this field. It will highlight existing processes that can be adopted and leveraged to deliver the safe outcomes, because effective process safety leadership is a key ingredient for a successful transition to a hydrogen economy.

Lessons learned from a fire at a Hydrogen 200 bar filling station

Conference

Center for Hydrogen Safety Conference Europe

Year

2023

Proceeding

2023 Center for Hydrogen Safety Europe Conference

Group

General Submissions

Session

Incidents and Lessons Learned

Time

Wednesday, May 10, 2023 - 10:15am to 10:30am

Authors

Ruban, S. - Presenter

Sharing lessons learned is a priority to build safety standards and culture among the hydrogen ecosystem. On the 30th of July 2020 at 5:15 pm, a high pressure leak was heard by two operators in a hydrogen filling station in France coming from a filling station. A fire broke out on a 200 bar hydrogen semi-trailer (composed of 4 bundles of 60 cylinders) being under filling operation at 178 bar. There were no injuries but the material consequences of the fire were severe: 11 cylinders within one bundle bursted and two adjacent hydrogen trailers partially burnt.

Emergency organization plans and the â€œimportant for safety elementsâ€ worked efficiently. The main side effects, some metal parts from the frame steel plates projected, remained inside the site, without any further domino effect.

The root cause analysis showed that most probably a small leak on fitting caused a rupture of a pigtail which induced a hydrogen jet fire, igniting the tyres and causing the cylinders to burst.
Lessons learned concerns the trailer design and maintenance and recommendations for new filling centers design.

Hydrogen Safety in Water Electrolysers

Wednesday, May 10, 2023 - 11:30am to 11:45am

Authors

Latsuzbaia, R. - Presenter

Known for its versatility and adaptability, hydrogen is seen as the key to reducing carbon emissions in the world's most polluting industries. Hydrogen itself is non-toxic, but it presents unique safety concerns in both production facilities and the infrastructure to distribute and utilize it. The production of hydrogen via electrolysis possesses unique safety challenges, due to co-production of oxygen and hydrogen in a two compartments of a single reactor separated by a thin membrane. Potential leakage through membrane or its rupture can rapidly cause formation of an explosive hydrogen-oxygen mixture inside of the reactor. Finding suitable and cost-effective solutions to mitigate the explosion hazard becomes especially more challenging for larger scale electrolyser plants. Various solutions have been proposed by academia and industry to mitigate such risks.

In the present work, we will discuss the hydrogen related hazards in electrolysers and will review available solutions proposed by academia and industry to mitigate the risks.

IEA TCP Task 43- Subtask Safety Distances: State of the Art

Tuesday, May 9, 2023 - 10:15am to 10:30am

Authors

Vyazmina, E. - Presenter, AIR LIQUIDE

Phillips, L.

Chang, R.

The large deployment of hydrogen technologies for new applications such as heat, power, transport and other emerging industrial utilizations is essential to meet targets for CO2 reduction. However, this will lead to the increasing density of hydrogen installations in a proximity to the population which have to handle hydrogen technologies.

Local regulations differ and provide different prescriptions for safety and/or separation distances in different geographies.

The purpose of this work is to give an insight on different methodologies and recommendations developed for hydrogen (mainly) risk management and consequences assessment of accidental scenarios.

The first objectives is to review available methodologies and to identify the divergent points on the methodology. For this purpose, a survey has been launched to obtain the needed inputs from the subtask participants.

The current work presents the outcomes of this survey, highlighting the gaps and suggesting the prioritization of the actions to take to bridge these gaps.

The Hygrid Project - a Hydrogen Blending Demonstration in a Gas Distribution Network - Design Considerations for Safe Operation

Thursday, May 11, 2023 - 11:20am to 11:35am

Authors

Kileti, P. - Presenter, National Grid

Cavanagh, C. - Presenter, National Grid

Schneider-Moran, T.

Bevacqua, R.

Wolf, J.

To meet the goals for decarbonization around the world and in the US, gas distribution utilities are developing plans to blend green hydrogen into their networks. Utility companies are supporting research, preparing for the transition, and implementing demonstration projects. These projects will produce store, and blend hydrogen using electrolysis powered with renewable electricity either delivered or produced local. National Grid and the Town of Hempstead, NY are developing a novel demonstration project that includes the expansion of an existing hydrogen vehicle fueling station to both upgrade its fueling performance in accordance with SAE standards and to expand its capacity. The additional capacity will be used for blending into an adjacent gas distribution network that was recently upgraded with HDPE pipelines. The system will be a single facility that will include a combination of vehicle fueling with pure compressed hydrogen, at or above 350 Bar, and blending up to 20% hydrogen into a 60 psig natural gas distribution system for delivery to residential buildings. Both systems are expected to operate simultaneously at times. Each application has different performance requirements such as utilization profiles, storage requirements and hydrogen purity standards. The potential for offsite impacts, such as emissions as well as the safety profile are of paramount concern in the design of the systems. The design will limit noise, discharges of gases and water and also to minimize the risk level while both uses are partially subject to different industry standards. The safety principles and the current or anticipated codes and standards to be applied at this facility as well as the application of the National Gridâ€™s process safety program and the unique aspects to the Qualitative Risk Assessment will be discussed. The safety features anticipated at this facility will also be discussed including the various fault detection, alarm and notification functions

Comparison of a Prescriptive and a Performance-Based Approach for Fire and Explosion Risk Assessment of Hydrogen Plant

Thursday, May 11, 2023 - 3:10pm to 3:25pm

Authors

Deckers, X. - Presenter

Kamaliseresht, K. - Presenter

For fire safety, new technologies come with a perception of risk. For hydrogen applications, it is important to refer to regulations and standards, e.g. NFPA 2 on Hydrogen Technologies Code [1]. Prescriptive-based provisions as a fixed set of rules in the standard are most often applied as they are easy to apply. It is important to understand that these standards also allow for performance-based provisions which lead to compliance with the goals and objectives of the standard. In this presentation we will explain the process and advantages of performing a risk-informed performance-based assessment [2] by:

1. Quantifying fire safety performance goals and objectives in a risk tolerance profile (both in fire event likelihoods and consequences) for property damage and business interruption.
2. Determine performance criteria in terms of the conditional probability of fire protection system success given a defined scenario, time-line, target threshold, damage limits and risk tolerance.
3. Verify the risk and safety levels of existing or proposed fire protection design options in a quantitative manner.

Both prescriptive and performance-based options will be applied on a case-study and the advantages of both will be presented. A practical tool will be presented to limit the time required for quality control and allow for a large number of scenarios to be analyzed in an automated way.

[1] NFPA 2 on Hydrogen Technologies Code, 2023

[2] Risk-informed, Performance-Based Industrial Fire Protection, Thomas F. Barry, 2002

Engineering Assessment Process for Hydrogen Blending in Natural Gas Transmission

Thursday, May 11, 2023 - 11:35am to 11:50am

Authors

Vonau, B. - Presenter

Theme:

TC Energy is committed to "zero is real" in every facet of our business. Information sharing and industry networking related to safety is a critical strategy in achieving our goals as we begin our journey in the hydrogen development space.

Summary and Purpose:

This presentation will review TC Energy's framework for pursuing a pilot project that blends Hydrogen into an existing natural gas transmission lateral. This will be an interactive session providing a suggested framework for other member organizations to consider and to obtain feedback on how TC Energy can optimize the framework moving forward.

Content and Description:

1) Pilot goals and desired outcomes - this section will provide insight into why TC Energy is pursuing a blend pilot and what we hope to achieve from the pilot

2) Lateral screening and prioritization - this section will provide insight into how laterals were prioritized to determine top candidates for H2 blending pilot including variables analyzed

3) System overview - this section will provide a high-level overview of the system that TC Energy is considering including blending scope and interconnect considerations

4) High level risk screening - this section will provide an overview into TC Energy's risk assessment process including top risks and bow-tie diagram results

5) Regulatory framework - this section will provide insight into TC Energy's interpretation of the regulatory framework which governs this conversion including compliance planning

6) Engineering assessments (EA) - this section will provide insight into the EA framework that TC Energy has adopted to identify and evaluate the hazards that may result from the service change including proposed action items

7) Closeout plan and next steps - this section will provide guidance related to next steps required to deliver the project including engineering deliverables, operational readiness planning, public engagement and operational readiness.

Liquid Hydrogen Maritime Transportation

Thursday, May 11, 2023 - 2:00pm to 2:15pm

Authors

Boulland, J. - Presenter

Transportation of hydrogen in bulk in liquid form in large volume in maritime transportation has gained interest with the development of the production of hydrogen.

Today, from a regulatory perspective, this activity of shipping is not regulated at the international level (through the International Maritime Organization (IMO)), although the IMO has issued in 2016 Interim Recommendations on the carriage of liquefied hydrogen in bulk that specifies requirements to be followed as an initial basis. This IMO document is meant to be reviewed when more knowledge and information is acquired with concrete projects. This document would later serve as a basis for incorporation of â€œHydrogenâ€� in the main IMO Code dedicated to transportation of liquefied gases in bulk (the IMO â€œIGCâ€� Code).

Today, one Liquid Hydrogen carrier (ship carrying liquid hydrogen) has been constructed and is in operation, the â€œSuiso Frontierâ€� and a significant number of ship projects are being considered, at various phases of development.

The carriage of liquid hydrogen in bulk carries its own set of challenges. Based on initial knowledge acquired on projects, this paper will give insights on the general challenges associated with maritime transportation of liquid hydrogen in bulk and share on the regulatory landscape and what lies ahead.

This paper is proposed by Bureau Veritas Marine & Offshore, a leading classification society in the maritime business, involved in the certification of ship projects and associated equipment onboard.

Refining Hydrogen VCE Modeling Approaches

Thursday, May 11, 2023 - 3:25pm to 3:40pm

Authors

Diakow, P. - Presenter

Malik, D., Baker Engineering and Risk Consultants, Inc. (BakerRisk)

Hydrogen is a key energy carrier for modern society. The breaking of the hydrogen bonds within traditional hydrocarbon molecules has been the primary mode of energy utilization since the industrial revolution. An increased focus on â€œnet-zeroâ€� greenhouse gas emissions, specifically carbon dioxide and methane, has resulted in a global push for lower carbon energy vectors, including pure hydrogen. Accurately modeling the dispersion, fire and explosion hazards associated with new and existing hydrogen production, distribution and transportation networks, and consumption is a key component to the safe expansion of these networks.

BakerRisk performed a series of very lean hydrogen-air vapor cloud explosion (VCE) tests as part of an internal research effort. The goal of these tests was to better understand the VCE hazards associated with very lean hydrogen-air mixtures (â‰¤14% H_2). Flame speeds and blast loads were measured using high speed video and an array of dynamic pressure transducers. This paper discusses the test setup and test results, including a comparison with data from prior congested tests. The measured flame speeds are compared to those predicted using computational fluid dynamics (CFD) analysis and referenced to deflagration-to-detonation (DDT) criteria. Discussion regarding the application of these test results and uncongested turbulent jet test results to facility siting studies is also provided.

Hydrogen Safety Management – Essential Practices for Risk Management in Operational Facilities

Wednesday, May 10, 2023 - 11:45am to 12:00pm

Authors

Ramachandran, N. - Presenter, Chola MS Risk Services

Dandamudi, R. - Presenter

Garimella, V. S., Cholamandalam MS Risk Services

Swaminathan, R.

Mani, S., Cholamandalam MS Risk Services

Even though the thrust and focus to move the hydrogen production to green hydrogen and blue hydrogen is kicking off, the risk of handling high pressure and high volume of hydrogen remains unchanged. It will take time for the transition to happen, and it is also imperative that these facilities will be operational for near future particularly in developing countries. Considering that the most operational facilities were set up at least twenty years ago, it is always challenging to manage the risk during operations and maintenance. This paper elaborates the risk management themes (assessing the risk; conduct of operations; asset integrity practices; transportation safety; fire detection, prevention, and protection practices etc.) to present the major safety challenges and the essential safety practices for operational Hydrogen generation units.

The analysis has been carried out based on various risk consulting services (E.g., Process Hazard Analysis, Functional Safety Assessments, Design of fire prevention and protection systems, Process safety management implementation, etc.) provided by Cholamandalam MS Risk Services for traditional Hydrogen generation units (across industrial sectors like Petroleum Refining, Fertilizer, Methanol, Chlor alkali industries and Thermal power plants). The major safety challenges, and essential safety practices for each of the risk management theme presented in this paper have already been implemented in Asia and found effective in nature. The paper also touches upon the way forward for decommissioning the traditional Hydrogen generation units. The overall idea is to present practicable essential safety practices for traditional Hydrogen manufacturing facilities to manage risk effectively during operation phase.

Hydrogen Use for Decarbonized Aircrafts and Associated Safety Challenges

Tuesday, May 9, 2023 - 3:40pm to 3:55pm

Authors

Wolf, P. - Presenter

Toussaint, B. - Presenter

Mahon, A.

Gay, A. M.

Aviation is responsible for around 2.5% of global CO2 emissions, with most aircrafts powered by kerosene. In October 2022, the UNâ€™s International Civil Aviation Organization (ICAO) adopted the goal of reaching low carbon emissions by 2050. Anticipating this target, Airbus unveiled its ZEROe programme, targeting a decarbonized aircraft to enter in service by 2035, using hydrogen fuel. The transition to this energy vector raises many challenges in terms of design and operation concepts for both in-flight and ground conditions. These are exposed in this article. Reconciling hydrogen physical and chemical properties, aeronautical applications specifics and safety targets is an ambitious objective. Aircraft hydrogen systems will face severe environmental conditions and drastic vehicle dynamics while having to satisfy operational cycling along with ensuring adequate level of safety. Yet designing an efficient and safe hydrogen powered aircraft is only a part of the numerous subjects to tackle. Aircraft certification and standardization remain open and evolving topics that are critical towards safe designs. Wide uncertainties are still to be addressed regarding the hydrogen ecosystem, from production to distribution and ground infrastructures: airport operations, safety and regulatory framework, availability, among others. Beyond Airbus and the aeronautical industry, global commitment is key to the success of this journey towards decarbonized aviation.

Process Safety Management for the Hydrogen Industry

Wednesday, May 10, 2023 - 11:15am to 11:30am

Authors

Moore, D. - Presenter, AcuTech Consulting Group

The hydrogen industry as it builds out will need to maintain the highest of safety standards to justify the validity of the industry for its many applications. Process Safety Management (PSM) is a proven approach to managing risks of facilities handling highly hazardous chemicals and is the global benchmark for managing catastrophic risks of chemical-related operations. Producers, suppliers, facility operators, users, and their contractors and employees would all benefit from an enterprise approach to PSM voluntarily applied even if not required by regulation and in addition to industry codes and standards. Some operations may be subject to PSM regulations, which may vary depending on the country of operation and their regulatory frameworks. Others may operate in countries that do not have a PSM regulation.

Experience of over 30 years of PSM has shown that it has changed the way business is managed and is deeply integrated into day-to-day operations. This paper will explain the application of PSM to hydrogen operations in manufacturing of hydrogen related products, the production of hydrogen, and the use of hydrogen as a fuel. Lessons learned from these applications will be explained to show the value and benefits.

Despite many accomplishments in preventing and managing catastrophic losses, it is clear that PSM requires a discipline to be truly effective. We believe that breakthroughs to a highly reliable and safe operation are only possible with PSM being oriented toward â€œprocess safety excellenceâ€?, including modern frameworks such as the AIChE CCPS Risk-Based Process Safety model, and then an emphasis on internal compliance to standards and practices.

To achieve process safety management excellence requires Conduct of Operations (COD) and Operational Discipline (OD). Formalizing COD/OD makes the difference between good and great control of process safety risks. Leadership in process safety requires clear direction, clear communications, and disciplined execution. This allows for improved process safety implementation including:

- auditable, documented compliance to standards
- more collaborative execution of key process safety activities
- improved communications
- improved process safety culture
- and many other benefits

Topics will include:

- The importance of global PSM application for the hydrogen industry
- Examples of successful application of PSM to the hydrogen industry
- The AIChE CCPS Risk-Based Process safety framework
- A model of COD/OD for process safety management
- Expected value and improvement potential

This paper will be particularly useful for companies who are producers or users of hydrogen to make them aware of the advantages of a formal management system for PSM. It will also benefit those who practice PSM but want breakthroughs in process safety performance and who recognize the leadership skills and efforts required.

Leak detection for hydrogen gas: atmospheric dispersion and optimal sensor placement

Tuesday, May 9, 2023 - 2:20pm to 2:35pm

Authors

Paleja, R. - Presenter

It is possible to store large quantities of energy in the form of hydrogen gas (H2), which can be created from water using electricity from renewable sources. Combustion of H2 produces only water vapour. For these reasons, hydrogen is seen as a valuable route to assist economies reach net zero emissions ambitions. The UK aims to deliver 5 GW of hydrogen production by 2030, while the EU has a 40 GW target.
However, hydrogen is a highly reactive and explosive fuel, prone to leakage from storage and processing equipment, posing a safety hazard. Moreover, atmospheric hydrogen gas has a global warming potential of around 10 times larger than that of CO2 over a 100-year horizon according to the UK government studies. Leak detection of atmospheric H2 is therefore of considerable importance. In this presentation, we consider the problem of optimal sensor placement for leak detection with different spot sensors technologies, using numerical studies to minimise relevant loss criteria. We describe the atmospheric dispersion of H2 leaking from some source location using different models, including a Gaussian Plume for steady-state advection-diffusion, and a jet model for gas release from a pressurised system.

The presentation also includes a wider discussion of how statistical modelling can help facilitate a safe, reliable and efficient hydrogen economy.

Evaluation of Safety Distances for Hydrogen Refueling Stations

Tuesday, May 9, 2023 - 12:05pm to 12:20pm

Authors

Truchot, B. - Presenter

Daubech, J., Institut National de l'Environnement Industriel et des Risques (INERIS),

Due to the strong increase in the hydrogen use for mobility, hydrogen refueling station (HRS) development is currently a major issue. Having a safe design of those HRS is crucial to prevent any major accident during the deployment of this technology. To accompany this safe development, a generic safety study was achieved. Based on the main scenarios identified in the risk analysis, associated safety distances were evaluated. Those scenarios include mainly tank burst, jet fire, jet explosion and explosion in container. To meet the objective of having relevant distances for all those scenarios, different approaches were used from simple correlation to CFD model.

Several leak diameters, from 10% to full line rupture, and pressures were considered, for both confined and atmospheric leakages. For confined leak, a specific model was applied to highlight the influence of the mixture concentration and turbulence, but also to consider the opening part of the structure, container typically.

Atmospheric dispersion of pressurized hydrogen was modelled using a computational fluid dynamic (CFD) code based on the large eddy simulation turbulence closure. Such an approach is essential to show the strong unsteady effect in such a release. The CFD model also enables ones to estimate the flammable mass in the case of impinging jets. For free jets, comparison between the CFD model, an integral approach and experimental data was proceed to ensure that all approaches are in accordance.

CFD was typically used, not only for jet dispersion modelling, but also to evaluate whether a surrounding fire could induce a tank burst, considering potential thermal barriers, based on the applied heat fluxes on the skin of the tank.

Multhyfuel Project for a Safe Implementation of Hydrogen Refuelling Station in a Multifuel Context : Midterm Achievements

Tuesday, May 9, 2023 - 11:30am to 11:50am

Authors

Sébastien, Q. - Presenter

The MultHyFuel Project, funded by the Clean Hydrogen Partnership, aims to achieve the effective and safe deployment of hydrogen as a carbon-neutral fuel, by developing a common strategy for implementing Hydrogen Refuelling Stations (HRS) in a multifuel context. The project contributes to the harmonisation of existing regulations, codes and standards (RCS) by generating practical, theoretical and experimental data related to HRS.

This presentation describes the goals and expected results of the project with the methodology followed by the project through the description of the different steps, work packages and the preliminary results of some activities already achieved such as the states of the art on technologies and risk assessment (preliminary and detailed).

The main results presented are the following :

- Three configurations of HRS in multi fuel context studied with different modes of supply of hydrogen (high-pressure storage (trailers or bundles), hydrogen production by on-site electrolysis, and stationary liquid hydrogen storage) and different operating conditions of the dispenser - i.e. flow and pressure - but only dispensing compressed gaseous hydrogen.
- A previous H2 and conventional fuel facility incident review to formulate lessons learned
- Identification of critical scenarios on H2 dispensers
- Preliminary recommendations of safety barriers for H2 dispensers

Finally, the experimentation plan for hydrogen dispenser and components is presented to complete the project and establish best practices for the safe implementation of HRS in a multi fuel context.

Safety Considerations in Cryogenic Liquid Hydrogen Bunkering

Thursday, May 11, 2023 - 1:45pm to 2:00pm

Authors

Wen, J. - Presenter, University of Warwick

The presentation aims to analyse safety considerations concerning the transfer of cryogenic liquid hydrogen (LH2). It will cover all three applications related to LH2 transfer, including filling a stationary tank, bunkering of ships or refuelling a truck. The relevant loss of containment (LOC) scenarios will be analysed. This will be followed by brief summary of the available large-scale test data, which provide useful insight while still leaving gaps in the understanding of some safety critical LOC scenarios. Examples of these include hydrogen jet fire attacks on LH2 tanks, effects of tank insulations, LH2 spill over water of different salinity and temperature, the propensity to rapid phase transition and ignition, both could result in semi-confined explosions between ship and harbour. The need for further large-scale tests to fill these safety critical gaps will be highlighted.

The presentation will then outline a framework for risk and consequence analysis using established risk assessment methods and computational fluid dynamics (CFD) techniques, incorporating consideration for barrier walls and safety distances.

Safety-related activities in the Clean Hydrogen Partnership

Tuesday, May 9, 2023 - 9:15am to 9:45am

Authors

Garcia Hombrados, A. - Presenter

The Clean Hydrogen Partnership (as per its legal name Clean Hydrogen Joint Undertaking) is a unique public-private partnership supporting research and innovation (R&I) activities in hydrogen technologies in Europe. The Clean Hydrogen JU is the successor of FCH 2 JU and was established in November 2021. With the mission of facilitating the transition to a greener EU society through the development of hydrogen technologies, the overall goal of the Clean Hydrogen JU is to support research and innovation (R&I) activities in the EU in clean hydrogen solutions and technologies, under the EUâ€™s funding Programme for research and innovation, Horizon Europe, and in synergy with other EU initiatives and Programmes. The EU supports the Clean Hydrogen JU with â‚¬1 billion euro for 2021-2027, complemented by at least an equivalent amount of private investment (from the partnership's private members of the partnership. Moreover, in the RePowerEU communication, the EU is additionally supporting the JU with 200 million euros for further developing the Hydrogen Valleys in Europe.

Safety is paramount for frictionless development, perception, acceptance of and trust in new technologies in modern society. Hydrogen is nowadays recognised as a key clean energy vector and hydrogen-based technologies are undergoing rapid expansion across multiple applications. This is a result of the credibility it has built up as a clean, flexible, versatile, reliable, and safe technology.

The JU's commitment to hydrogen safety is sound and the JU supports numerous activities focused on hydrogen safety, contributing to reach a high-level degree of hydrogen safety while contributing to building the confidence needed for widespread take-up and safe deployment of hydrogen and fuel cell technologies.

The Clean Hydrogen JU is addressing hydrogen safety-related aspects from different perspectives: in research and innovation projects, through the collaboration on safety aspects with the Joint Research Centre (JRC) of the European Commission and through the activities undertaken by European Hydrogen Safety Panel (EHSP). The presentation provides an overview of the hydrogen safety-related activities supported by the Clean Hydrogen JU programme for a safe rollout of the FCH technologies in Europe and beyond.

Mitigation of Electrostatic Ignition Hazards during the Bulk Loading and Fueling of Road Tankers, Railcars and Public Transportation Vehicles.

Thursday, May 11, 2023 - 1:15pm to 1:30pm

Authors

Rowe, J. - Presenter

O'Brien, M.

This presentation will aid engineers, operations managers and safety professionals in their understanding of the root causes of static electricity and why it can present as a potential ignition source during the transfer of hydrogen in bulk loading and fueling operations for road tankers, rail cars and public transportation vehicles. It will provide a general overview of industry guidance relevant to static electricity, including the requirements outlined in Part 1 of ISO 19880-1 â€œGaseous hydrogen â€" Fueling Stationsâ€, (2020). The presentation will also address aspects of the hazardous area product certification requirements for hydrogen gas atmospheres as per the IEC 60079 series of explosive atmosphere equipment standards. The presentation will finish with some practical examples of good static control practices that can be used to support SOPs designed to mitigate against the unwanted accumulation of static electricity on the transportation vehicle undergoing the hydrogen fueling operation.

Eliminating Hydrogen Fire in CCR Units

Wednesday, May 10, 2023 - 10:00am to 10:15am

Authors

Alharkan, M. - Presenter

Almutairi, M. - Presenter

In the propane dehydrogenation facility, the CCR unit is supplied by UOP. This unit is able to continuously regenerate catalysts based on a dedicated PLC control logic. The system operates automatic process valves to regulate and control the flow of the catalyst through the system, isolating and altering process atmospheres..

The Oleflex catalyst regeneration system requires catalyst transfer from hydrogen/ hydrocarbon atmosphere in reactors to the oxygen rich atmosphere in the regenerator and back again. This requires the valves to be operated at a high cycle normally about 8 times an hour.

Due to this high operation cycle, the gland packing tends to become loose leading to minor passing. And due to presence of hydrogen, there are chances of fires.

It was identified that high rate of packing leaks was observed in the Fisher V-Ball and Metso ball valve. These are rotary valves while compared to other linear valves. Both the valves have conventional packing design with nut and bolt gland tensioning using gland followers and lock nut. This is the conventional style used in general purpose valve design.

It was observed that the valve started to leak from the gland even after one month of torque.

Also, the packing material used is laminate graphite. Better sealing will be provided by PTFE. The

limitation is temperature should be less than 230 Deg Operations have confirmed that the operating.

temperature is well below 200 Deg C. Therefore, a reliable alternative would be to replace the valve with an enviro seal package with PTFE material.

Safe measures to mitigate the effects of a surrounding fire on gaseous hydrogen composite tube trailers.

Thursday, May 11, 2023 - 1:30pm to 1:45pm

Authors

Guerin, C. - Presenter

The growth of hydrogen demand for mobility leads to an increase in the number of GH2 trailers on the road.

Composite cylinders have been developed to increase the amount of hydrogen carried and higher pressure.

One of the major safety scenarios is a fire on a GH2 composite tube trailer which could be caused by a road accident or a large hydrogen leak, or a brake system fire.

When composite tubes are subjected to a surrounding fire, the mechanical performance of the multilayers and liner tightness degrades. Tubes will then leak or burst depending on pressure and temperature.

This enhances the need to depressurize the composite tubes in the event of a fire before degradation of the mechanical properties of tubes and burst. Due to the thermal insulation properties of the material, heat from the fire will not raise pressure in the cylinder significantly. Therefore, a pressure safety valve (opening in case of overpressure) is not suitable. A TPRD (Thermally activated Pressure Relief Device) is a right safety device to depressurize composite tubes in this case because TPRD opens automatically when its temperature threshold is reached (~110°C).

Recent road incidents showed the efficiency of TPRD to limit the consequences of a fire on GH2 composite trailers.

About hydrogen leak detection. Overview of existing devices and investigations on innovative technologies for safety considerations.

Tuesday, May 9, 2023 - 1:45pm to 2:00pm

Authors

Houssin, D. - Presenter

Hydrogen is an essential energy vector and it is required to master safety of H_2 energy on the whole value chain including new applications, new environments and new end-users.

Leak detection is one of the key mitigation barriers which avoids catastrophic consequences or limits the severity of hazardous scenarios. In most cases, an unintended release of gaseous or liquid hydrogen could lead to potential dangerous consequences such as fire or explosion. Other reasons for the early detection of hydrogen releases are environmental aspects, reliability of infrastructure, rationalized consumption, energy savings... The experience shows that nowadays there is no universal leak detection system which has high performances in different operation conditions and surrounding environment: detection equipment effective in confined areas can be totally inefficient in free field, in large buildings, in well ventilated spaces... Detection systems have to be smartly chosen and installed in accordance with the targeted application and their environment of use.

This communication shed a light on the main devices/detection techniques usable for hydrogen leak detection, highlighting advantages and drawbacks of such equipment. Emerging technologies will be introduced - as well - through on-the-field investigations aiming at assessing their usability and benefits.

AUTHOR INDEX

Alharkan, Manea 28
Almutairi, Majdi 28
Assunção, João Vitor 8
Bevacqua, Ron 13
Bouchemoua, Yanis 3
Boulland, Julien 16
Cavanagh, Christopher 13
Chang, Richard 12
Dandamudi, Raja 18
Daubech, J. 23
Deckers, Xavier 14
Diakow, Peter 17
Duchateau, Maxime 3
Esteves, Vinicius 8
Garimella, Venkata Sainath 18
Gay, Anaïs Marie 19
Guerin, Christophe 29
Hombrados, Alberto Garcia 26
Houssin, Deborah 30
Ibarreta, Alfonso 5
Jeghloul, Sara 3
Kamaliseresht, Khodadad 14
Kerin, Trish 9
Kileti, Pradheep 13
Mahon, Alban 19
Malik, Darren 17
Mani, Senthilraja 18
Moore, David 20
Morelli, Enrico 11
Myers, Timothy 5
O'Brien, Mike 27
Paleja, Rakesh 22
Pereira, Rafael 8
Perrodin, Florian 6
Phillips, Lee 12
Quilliou, Josué 3
Ramachandran, Nagarajan 18
Reuderink, Pim 1
Rowe, Jonathan 27
Ruban, Sidonie 10
Schneider-Moran, Tara 13
Sébastien, Quesnel 24
Swaminathan, Raman 18
Tchouvelev, Andrei V 2
Torrado, David 4
Toussaint, Beatrice 19
Truchot, Benjamin 23
Vejayan, Harmina 6
Vonau, Bradley 15
Vyazmina, Elena 12
Wechsung, Achim 5
Wehmuth, Bruno 8
Wen, Jennifer 25
Wennesjö, Martina 7
Wolf, John 13
Wolf, Pierre 19